The Book of Physics Anagrams and Puzzles

- Book 1

Independent Publishing Network

Published in 2024 by Independent Publishing Network

An imprint of Independent Publishing Network

The Independent Publishing Network is operated by International Barcodes Limited in partnership with Independent Publishing Network Ltd.
email: contact@bookisbn.org.uk

International Barcodes Limited
2C Handley Ave, Devonport, Auckland 0624, New Zealand
NZBN: 9429030122956

Independent Publishing Network Ltd
Mailing address – MB #1869, PO BOX 480, Sevenoaks TN13 9JY UK
Registered Office – 71-75 Shelton Street, Covent Garden, London, WC2H 9JQ, UK
Company Number 11541223

This is an updated edition of *Book of Physics Anagrams and Puzzles – Book1*, first published February 4th, 2024.

ISBN

Paperback- 978-1-80517-425-7

Book of Physics Anagrams, Crossword puzzles, Word-searches and Mind-maps

Research has shown that learners improve their problem-solving skills by working through puzzles, even from infancy. Children love working their way through jigsaw puzzles. As they watch the big picture emerging, they develop a sense of fulfilment and accomplishments. It is amazing how much skills learners gain from **word-searches** and **puzzles**.

This book has been developed from working closely with learners in the secondary school. The success of engaging the learners in activities that enthuse most has been exhilarating for me as a teacher. I decided to put these into a book form after a talented learner of honour disparaged the first activity in the chain (the **word-search**). Learner felt they could never learn anything from this type of activity. But as the chain of activities grew, learner became so engrossed in them. At the end of the lesson, learner felt that there was so much to be gained even by a class of very diverse abilities.

The chain of activities begins with an **anagram** of key words, the progresses to a **word-search** after which it takes the learner into the world of **crossword** puzzle. The last link in the chain is the **mind-map** activities. All are geared to stimulate the thinking of the learner, as well as get them familiar with the vocabulary needed for higher-level responses.

Many learners have found these activities quite helpful in preparing for their in-school tests, external examination, and as aids to understanding their homework. Teachers also could use the bank of activities as "scaffolding" to help learners grasp the abstract concept in Physics.

This book covers the broad topics common to the various examination boards in the UK, and is adaptable to international flavours. As such, it can be used by all learners irrespective of the examination board featured in their schools.

I urge the learners using this book never to quit at the first sign of challenge. You will only get better as you practice these questions. Also, work through the questions with your friends before you look at the solutions at the back of the book.

As a teacher of many years, I have seen learners progress in attainment because of their hard-work at varying activities, and on to achieving excellent results. I hope you find the same delight in using the resources in this book as additional aid in your preparation.

Fol Marc

Table of Contents

For each sub-topic there would be

- **Summary**
- **Anagram**
- **Wordsearch**
- **Crossword puzzle**
- **Mind map**

- Summary -

1. Matter: *The particle model*

People of ancient times wondered if everyday matter were made of tiny building blocks. The ancient Greeks called these atoms, which means indivisible. Some 120 years ago, those who practice natural philosophy, or science, became interested in atoms and how they could be moved around, although, even as at then, most felt that there could be nothing of the sort like atoms. Progressively, scientists began to develop the idea of atoms as they conducted more investigations. Ideas that did not conform with observations were discarded for new ones. More scientists got involved in the study of the world around them. They began to embrace these new discoveries; that everyday matter, visible and invisible, were made of atoms ; that atoms were made of tinier particles, although most of it was empty space; that the charge particles are separated in the atom, held in place by an attractive force, that the central part contains the positive charges; that the negative charges would have to whizz around the centre to stay attracted; that most of the mass of the atom is in its centre; that most atoms also have uncharged particles at the centre.

- Anagram -

1. Matter: *The particle model*

Unscramble the key words below:

1. oamt   =>   _ _ _ _

2. icomat eldom   =>   _ _ _ _ _ _ _ _ _ _

3. lamsounoa   =>   _ _ _ _ _ _ _ _ _

4. ydniets   =>   _ _ _ _ _ _ _

5. gyeenr lveel   =>   _ _ _ _ _ _ _ _ _ _ _

6. truenal   =>   _ _ _ _ _ _ _

7. cneslu   =>   _ _ _ _ _ _

8. rtiob   =>   _ _ _ _ _

9. ricptlae odlem =>   _ _ _ _ _ _ _ _ _ _ _ _ _

10. lslhe   =>   _ _ _ _ _

11. triao   =>   _ _ _ _ _

12. elverita   =>   _ _ _ _ _ _ _ _

13. ssma   =>   _ _ _ _

14. grceha   =>   _ _ _ _ _ _

15. oemvlu   =>   _ _ _ _ _ _

1. Matter: *The particle model*

Search for the unscrambled key words in the word-search below. Note down the time it took you to find all the words.

```
S W S U O W X T O O W D N T J
P S Q V W P L B N V I B H Q E
V G A O I X B L Z M O T A P A
G V K M W S U R E N V G A O T
U R N C Z U E E U H P Q X R O
C Y Z U H L N C A H S F V T M
L B O E A E L M K Q S N O I I
E B I T U E C M N Z U W L B C
M I I T U P T M A P N Q U R M
D V R S Z T C H A R G E M O O
E A X T D E N S I T Y F E Q D
L P A R T I C L E M O D E L E
U J D A J S U O L A M O N A L
E N E R G Y L E V E L F R X I
V B Q Z O G O X N I T B Z R Y
```

Duration:

1. Matter: *The particle model*

- Crossword -

Use the clues to fill out the words for the crossword puzzle

Across:

3. net charge of an atom

4. negative for electrons

5. in comparison

9. Unusual behaviour; not fitting pattern

11. places electrons can be from nucleus

12. levels indicating potential energy (2)

14. The path of electrons around nucleus

Down:

1. A view of particle arrangement (2)

2. Length **x** width **x** height

6. smallest unit of matter

7. total amount of matter

8. central part of an atom

10. ratio of mass to volume

13. proportional relation

Mind-map template: *The particle model*

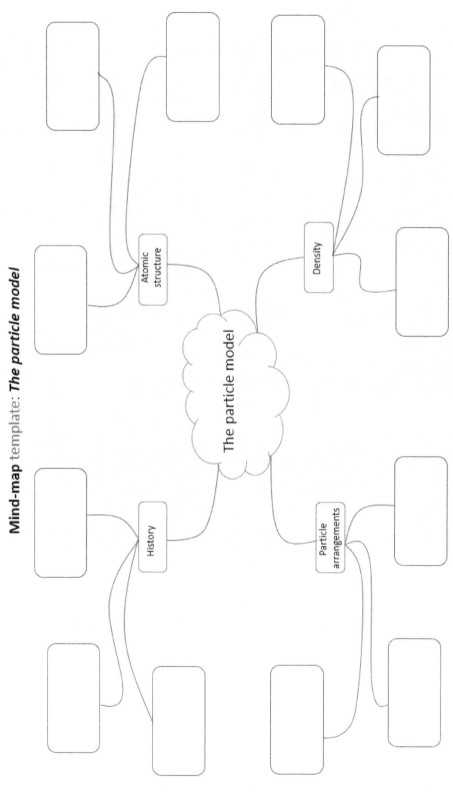

The particle model

Atomic structure

Density

History

Particle arrangements

Suggestion: *Use a pencil to complete the mind-map in case you change your mind.*

2. Matter: *Changes of state*

It is generally accepted that there are three states of matter: solid, liquid and gas (some propose plasma as a separate state of matter). Even at room temperature some substances melt when heated or freeze when cooled. Learners should be able to explain the changes of state in terms of the changes to the arrangement of particles in substances, as well as the forces holding these particles in place. They should be able to recall and correctly use the concept of temperature. They need to be familiar with the three main scales in use and begin to make conversions between the Celsius and Kelvin scales. Going further, they should be able to use the concept of internal energy (change in energy of the "unseen" particles), and link this to the kinetic energy (change in the motion of the particles), and the potential energy (changes to the intermolecular forces or bonds) of these particles. They would need to understand the link between the specific heat capacity of materials and the increase in the kinetic energy, and that of the specific latent heat and its connection to the specific latent heat of substances, as well as explaining related graphs.

2. Matter: *Changes of state*

Unscramble the key words below:

1. aleutsob alecs => _ _ _ _ _ _ _ _ _ _ _ _ _

2. glibion => _ _ _ _ _ _ _

3. dsnbo => _ _ _ _ _

4. ghcaen fo steta => _ _ _ _ _ _ _ _ _ _ _ _ _

5. nanosecondit => _ _ _ _ _ _ _ _ _ _ _

6. innovatesroc => _ _ _ _ _ _ _ _ _ _ _

7. regeny ernstfar => _ _ _ _ _ _ _ _ _ _ _ _

8. atropineova => _ _ _ _ _ _ _ _ _ _

9. nigerfez => _ _ _ _ _ _ _ _

10. corallinemuter => _ _ _ _ _ _ _ _ _ _ _ _ _

11. antrenil egyenr => _ _ _ _ _ _ _ _ _ _ _ _ _ _

12. inkteci yngeer => _ _ _ _ _ _ _ _ _ _ _ _ _

13. tingelm => _ _ _ _ _ _ _ _

14. latinopet neryeg => _ _ _ _ _ _ _ _ _ _ _ _ _ _ _

15. athe => _ _ _ _

16. talent => _ _ _ _ _ _

17. limousinbat => _ _ _ _ _ _ _ _ _ _ _

18. attemperrue => _ _ _ _ _ _ _ _ _ _

2. Matter: *Changes of state*

- *Word-search* -

Search for the unscrambled key words in the grid of words below. Write the time it took you to find all the words at the bottom of the page.

```
B J F F Q Y F N W G I W H T V X R M G Y
Z C I S E P V D Q G T X F T B E N U N Q
S J L J R T B O I L I N G N F T O O I S
K R U X U B A W N H X I E S L D I R Z W
V W N I T M M T O Q Q H N T A T T U E C
B V O Y A W B O S J D A G R A S A K E B
P M O K R L I U E F R R H V N L S I R N
F R G X E D A C U T O T R R N K N I F R
Y X H R P S Q I Y Q Q E B K O K E W Q B
F O A F M K D G T E S J G P I F D F X D
W K T A E H R N I N F C H N T N N V E F
V K I N T E R M O L E C U L A R O R K B
S T N N N W T C E B U T C F M H C C S D
I M T E E P Q U L A W O O B I M C X C V
Z T E S S T J I L S O Z E P L Z Y G J U
V V R L X E I P Y O U O U Z B T S G E T
W X N C T G O C F J S K Q X U K I I B C
Y U A N K I T W G N D B W W S D K B F C
A C L D T P N U N O I T A R O P A V E Y
G J T A K Y W G O I R Y T A S J Z O P K
```

Duration:

2. Matter: *Changes of state*
- Crossword -

Use the clues to fill out the words for the crossword puzzle

Across:

6. Energy stored in (intermolecular)bonds
8. Hidden (energy)
9. change: solid to liquid
13. Gas ⇔ liquid ⇔ solid
15. change: solid to gas
16. change: gas to liquid
17. Energy due to movement

Down:

1. change: liquid to solid
2. change: liquid to gas (all temp.)
3. change: liquid to gas (1 temp.)
4. total energy of particles
5. Energy due to temp. difference
7. Bond between molecules
10. In comparison
11. Total amount of matter
12. Keeping things the same
14. Kelvin temperature scale

Mind-map template: *Changes of state*

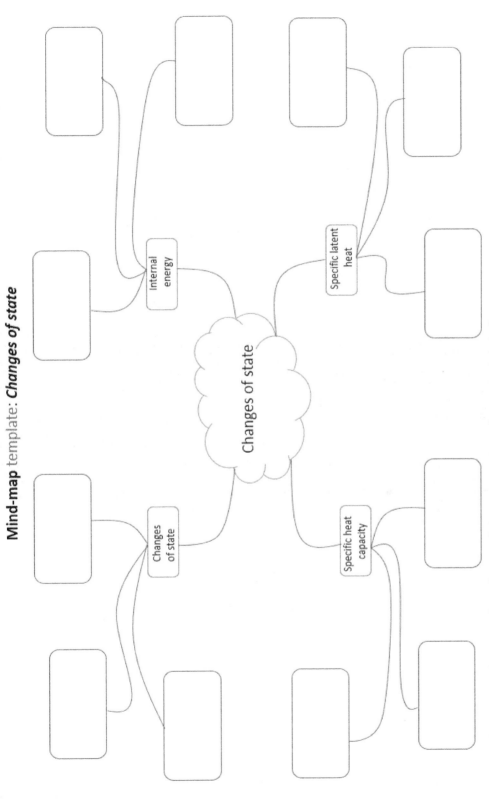

Changes of state

Internal energy

Specific latent heat

Changes of state

Specific heat capacity

3. Matter: *Pressure*

Water pressure is easy for learners to relate to. When they feel a balloon with water, they feel its pressure. And most can relate it to the weight and density of the liquid. They need to recall the relationship between pressure, weight, and area, and be able to explain how this explains variation in pressure when any of the two quantities are changed. While most learners can easily visualise the "push" back from closely packed particles of liquids and solids, they need to be able to describe how movement of gas molecules leads to pressure by these sparsely distributed particles. Learners also need to be confident they can explain why pressure increases with increasing temperature for the same mass of gas.

Many learners are surprised when they realise the magnitude of the pressure the earth's atmosphere exerts at the surface of the earth. While it is in theory differentiated into layers at various heights, they need to recall that the average density is roughly the same in all layers. Learners also need to be able to relate the incompressibility of liquids under pressure to their use in transmitting a force through the liquid and using this in force multipliers.

3. Matter: *Pressure*

Unscramble the key words below:

1. portechimas sprueres=> _ _ _ _ _ _ _ _ _ _ _ _ _ _ _ _ _ _ _

2. sga erseuspr   =>   _ _ _ _ _ _ _ _ _ _ _

3. prosdecems   =>   _ _ _ _ _ _ _ _ _ _

4. insobrietyslimpic   =>   _ _ _ _ _ _ _ _ _ _ _ _ _ _ _ _ _

5. silveryen propanoltrio=> _ _ _ _ _ _ _ _ _ _ _ _ _ _ _ _ _ _ _ _

6. lamnor   =>   _ _ _ _ _ _

7. saclap   =>   _ _ _ _ _ _

8. lordmany   =>   _ _ _ _ _ _ _ _ _

9. euterpetram   =>   _ _ _ _ _ _ _ _

10. ruthstup   =>   _ _ _ _ _ _ _ _ _ _ _ _ _ _

11. olmuev   =>   _ _ _ _ _ _

3. Matter: *Pressure*

Search for the unscrambled key words in the grid of words below. Write the time it took you to find all the words at the bottom of the page.

```
I E X W C X N X X R J K D U U U C M W F A
F N V E M J K N V G C V Q O V S S X L T Z
F Y V R D G D L E L A F C L F F I G E X O
F P G E Z A K W R U A W Z N E I Q M B T S
B R U Q R H S C S S I S Q M E T P D X J A
M A K I J S V P T M N R U Q J E A M J O Y
I L N C V Z E A P G C L Y G R C S L E I J
F F P F K H Y L Y N O R M A L T C P V U V
U D M P A A O Z Y V M D T S Q O A C E A D
K P T N B T K R M P P U E A M H L X W I M
T K X Y L M O D N A R A T P Z X N E I Q G
P L W W D O C S F E E O R H G A G F T F G
G N L A Q S Z K M J S E P Z Z C M Y Y I Q
A D Z Z W P L B Y M S H A O L U G V X L S
V G R X J H T D Z S I T G W R P A K F Q G
C Z A E E E S O E O B X E P T T B T Z K M
Z H P L S R H D L Y I B E V V H I B T H R
H Q D V I I K E I D L R Q A G R J O G M F
L W A M I C E J A Z I C G C O U C I N S P
P E Y Y F X Z Y Q E T H J L J S P R P A G
H N A J F F R J P F Y T V Z N T C K Q O L
```

Duration:

3. Matter: *Pressure*

- Crossword -

Use the clues to fill out the words for the crossword puzzle

Across:
2. force in fluids due to pressure difference
6. unit of pressure
7. average kinetic energy of particles
10. at right angles to surface
11. pressure versus volume

Down:
1. gases mixture above us
3. not possible with liquids
4. state with van der waal forces
5. variation with pressure (⇧⇩)
8. particle movement in gases
9. Brings gas particles closer

Mind-map template: *Pressure*

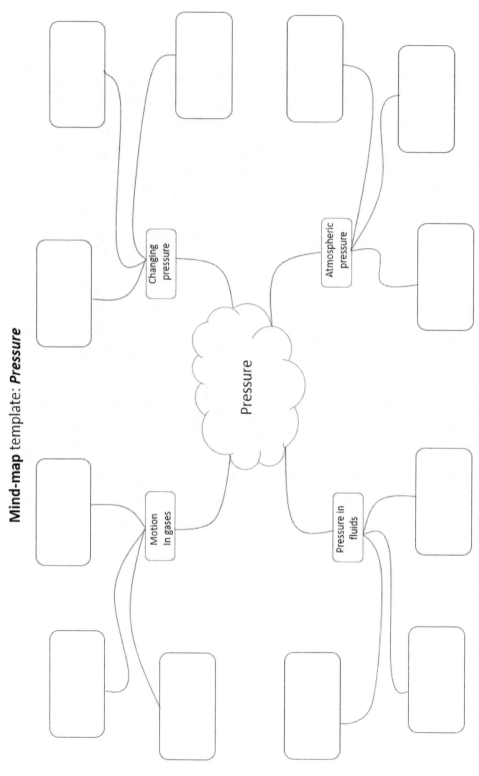

Central: Pressure

Branches: Motion in gases, Changing pressure, Atmospheric pressure, Pressure in fluids

Suggestion: *Use a pencil to complete the mind-map in case you change your mind.*

The Big Picture

The last three topics form a single unit. Can you see the inter-relationship?

Write (or sketch) below three key ideas that link the three sub-units

We also probe the world around us with big issues/questions

Write down three ideas/questions you would like to follow up on

1.

2.

3.

4. Forces: *Motion*

Motion continues to fascinate learners just as it had done with past generations. In describing motion Learners need to understand the differences between the terms that are used. Like the difference between speed and velocity, distance and displacement, and other measurements used in the description of motion. Learners need to be able to characterise these quantities either as scalars or vectors and recognise how this designation would affect their cumulative effects. "Practice makes perfect" is the way the adage goes. Just as a skilful tennis player will not relegate routine training to secondary place, so learners need to assign top priority to practising their calculations. Important still is the need to learn and recall equations especially those associated with motion, and that are not given in the examinations' supplementary materials. Graphs are excellent tools in visualising motion. Learners should be able to interpret, describe, and draw graphs of motion.

4. Forces: *Motion*

Unscramble the key words featured below:

1. learoctecani   =>   _ _ _ _ _ _ _ _ _ _ _

2. cartesianrise   =>   _ _ _ _ _ _ _ _ _ _ _ _ _

3. redseapeaveg   =>   _ _ _ _ _ _ _ _ _ _ _ _

4. clonedeartie   =>   _ _ _ _ _ _ _ _ _ _ _ _

5. sedimentclap   =>   _ _ _ _ _ _ _ _ _ _ _ _

6. nicestad   =>   _ _ _ _ _ _ _ _

7. rigatend   =>   _ _ _ _ _ _ _ _

8. tear fo ngeach   =>   _ _ _ _   _ _   _ _ _ _ _ _

9. rascal   =>   _ _ _ _ _ _

10. edeps   =>   _ _ _ _ _

11. nettnag   =>   _ _ _ _ _ _

12. finorum nimtoo   =>   _ _ _ _ _ _ _   _ _ _ _ _ _

13. loveytic   =>   _ _ _ _ _ _ _ _

4. Forces: *Motion*

- Wordsearch -

Search for the unscrambled key words in the grid of words below. Write the time it took you to find all the words at the bottom of the page.

```
W V L A U Y J P M S L N H T H J P Y
D D Y C K K F V G C I P P K H O W A
Y R W C R R C Y J A R G C E E F L T
S A P E K S Q W X L A J N N K K T G
F D J L T N E G N A T D A N C S V Z
L D E E P S E G A R E V A J U A T N
A Q M R F F Z V U C O L I Z T U Z K
G C R A G S Y D E B F E R X M A T I
C R I T T X D L F Y C O R H P H N L
L M M I S Q E N B N H W E K F L B H
U Z B O Q R V E A L A Y S P E E D T
S I O N A F A T J O N Q I L K W U N
L U P T G M S D W P G I S M M S F E
E H I D D I X E U O E Z T P W P R I
T O Y E D T T N E M E C A L P S I D
N N O I T O M M R O F I N U D K J A
F W Q F U U Q M S Y W C C I T F R R
Y D P Z K W Y T I C O L E V U C Q G
```

Duration:

4. Forces: *Motion*

- Crossword -

Use the clues to fill out the words for the crossword puzzle

Across:
4. 8(**Across**) divided by time
6. 5(**Down**) divided by time
7. Slowing down
8. Length in speed measurements
9. Drawn to touch curves for **10(Down)**
12. Opposes objects moving in air
13. Motion where **6(Across)** is constant

Down:
1. What gradient is
2. Length with direction
3. Direction is not important
5. 2(**Down**) divided by time
10. Rate of change
11. Size with direction

Mind-map template: *Motion*

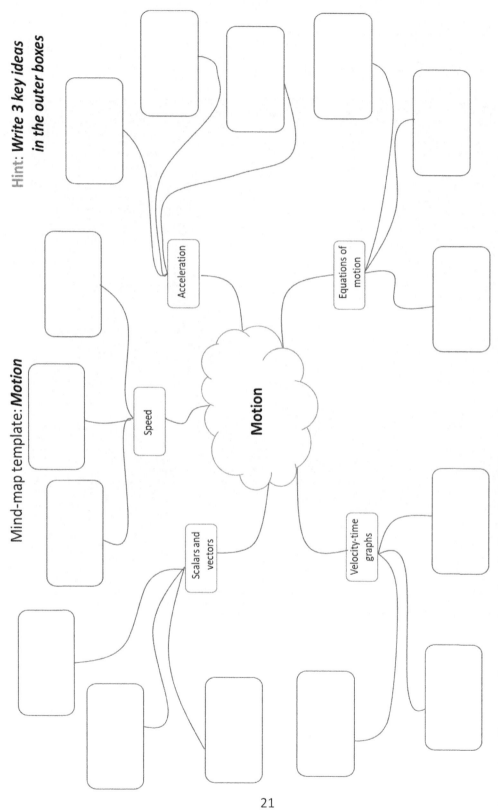

Motion

Speed

Acceleration

Equations of motion

Scalars and vectors

Velocity-time graphs

21

Suggestion: *Use a pencil to complete the mind-map in case you change your mind.*

5. Forces: *Newton's Laws of motion*

By empirical methods Isaac Newton tested and characterised three laws of motion. These laws are essential for learners to understand, interpret, and describe classical motion. The first, also referred to as the "law of inertia", explores and quantifies the relationship between forces and motion. The second, usually stated as "f=ma" describes the direct relationship between force and acceleration (which learners should relate to changes in speed, or direction even at constant speed). The third, the most familiar to non-scientists, is simply stated as "action and reaction are equal and opposite". Learners are expected to be able to determine the newton pair of forces using three attributes: they must be equal and opposite, same type of forces, and act on different objects. It is also essential that learners are able to relate these laws to the study of momentum, which then leads on to the design of safety features in when objects are in motion.

5. Forces: *Newton's Laws of motion*

Unscramble the key words below:

1. lucerpm ozne   =>   _ _ _ _ _ _ _   _ _ _ _

2. blueriiqmiu   =>   _ _ _ _ _ _ _ _ _ _ _

3. corfe   =>   _ _ _ _ _

4. elandbca   =>   _ _ _ _ _ _ _ _

5. optecomnet   =>   _ _ _ _ _ _ _ _

6. tcacnot   =>   _ _ _ _ _ _ _

7. lutearnts   =>   _ _ _ _ _ _ _ _ _

8. brdefyoe   =>   _ _ _ _ - _ _ _ _

9. antenir   =>   _ _ _ _ _ _ _ _

10. utmomnem   =>   _ _ _ _ _ _ _ _

11. oprew   =>   _ _ _ _ _

12. rokw   =>   _ _ _ _

5. Forces: *Newton's Laws of motion*

- Word-search -

Search for the unscrambled key words in the grid of words below. Write the time it took you to find all the words at the bottom of the page.

```
O Z U H H Z U Y K H W V E C T O R N J
L H C O N S E R V A T I O N E S E W V
S Y N Q G L V X H S N M E L W W V O G
N D C Q E R P R T D D N C V T Q A Q F
E Y H T E F A A O N O N C O N T A C T
Q J H W L L N I W P F H N J W E Z C J
U B O R A G V R M B N S P E E D A J K
I P U C E B D O U S N T I B I Y C W I
L O S N K P C D T B P G C K Y H R O N
I T T X M Z O K A S H K Q H W K U M E
B P C C V U T I R T B W Y V M N M Q T
R U U A T A T F A O R S J X E X P Z I
I J S J T R A N U M W X M Y M R L I C
U L H F E N F N E S B A L A N C E D E
M E U N I F O R M M O T I O N Z Z U N
J B I U G P R C S J O G X O J U O X E
S L U W O W C U C M E M E C G R N N R
L F W R P R E S U L T A N T P T E D G
F R E E B O D Y D I A G R A M V E C Y
```

Duration:

5. Forces: *Newton's Laws of motion*
- Crossword -

Use the clues to fill out the words for the crossword puzzle

Across:

1. Total remains the same

3. State of balanced forced

5. Only magnitude matters

6. Drawn on curves to calculate gradient

7. Rate of energy transfer

8. Forces that don't touch to act

11. Product of mass and velocity (m x v)

15. Unit of force

17. Measure of resistance to motion changes

18. Needs booth magnitude and direction

19. Energy absorbed by safety features

20. Motion governed by "suvat" equations

Down:

1. Safety feature delays crushing

2. Force due to pull of gravity

4. Diagram with arrows away from dot

9. Part of a resolved force

10. When net force is zero

12. Effective force, replacing multiples

13. Product of mass and acceleration

14. Rate of change of distance with time

16. Product of force and distance

Mind-map template: *Newton's Laws*

Hint: *Write 3 key ideas in the outer boxes*

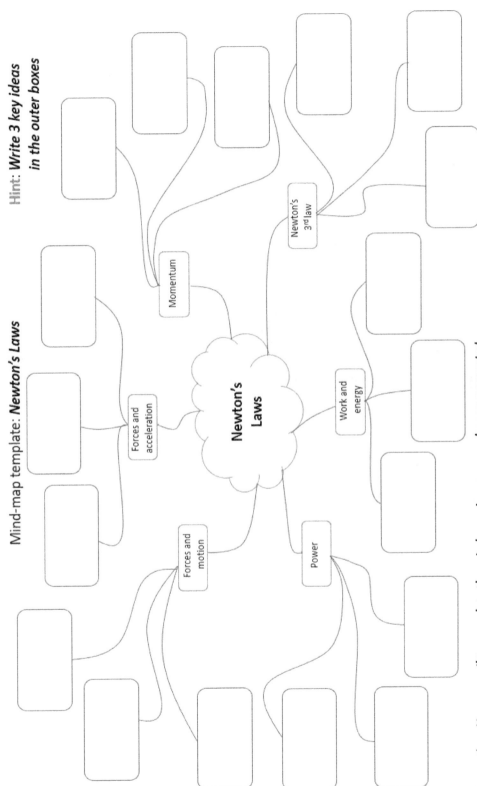

Newton's
Laws

Forces and
acceleration

Momentum

Newton's
3rd law

Work and
energy

Forces and
motion

Power

Suggestion: *Use a pencil to complete the mind-map in case you change your mind.*

6. Forces: *Forces in action*

Forces produce measurable effects, and motion that could be predicted (deterministic). Under this topic learners are introduced to different forces, the action they produced. They learn that when forces do work, they transfer energy, and that the total energy in any (closed) system is always conserved. Learners are introduced to mass as the quantification of matter in objects, and weight as the action of gravity on masses. The transformation from kinetic to gravitational potential energy with the associated calculations forms a central part of this unit. They spend some time examining Hooke's law relating force to extension produced in elastic materials, as well as in deformation of materials. A consideration of the application of force multipliers in our everyday living forms the conclusion of this unit.

6. Forces: *Forces in action*

Unscramble the key words featured below:

1. spoonericsm   => _ _ _ _ _ _ _ _ _ _ _

2. mordantefio   => _ _ _ _ _ _ _ _ _ _ _

3. scaleit   => _ _ _ _ _ _ _

4. ascilpt   => _ _ _ _ _ _ _

5. steinoxen   => _ _ _ _ _ _ _ _ _

6. erga   => _ _ _ _ _ _ _ _ _

7. erelv   => _ _ _ _ _

8. diycrulah   => _ _ _ _ _ _ _ _ _ _

9. eranil   => _ _ _ _ _ _

10. notemm   => _ _ _ _ _ _

11. optiv   => _ _ _ _ _

12. fitnessfs   => _ _ _ _ _ _ _ _ _

6. Forces: *Forces in action*
- Word-search -

Search for the unscrambled key words in the grid of words below. Write the time it took you to find all the words at the bottom of the page.

B T H A L C B N O I S N E T X E I
D U O G R Q W T T F I S W L Z T F
E E S V O D Z X Q E V G X V I S E
K F S J I F X Y A T Q P Z U G L O
F T E M I P Z J T R O F F E A M F
J M N U I N P B K X B B U S K N D
F G F F V E L H A R N Y T Q Z O U
T H F H N O F U G S M I P A E I H
E F I U A J O T V F C Q G E B S Y
P Y T D N O I T A M R O F E D S D
W M S B Y M Z X K A V A N M H E R
X I H Q X O R E Z X O B P K K R A
D D V K S M R E V E L H D D A P U
Z O I V F E I L I N E A R E A M L
D T G Q X N T H G I E W G Y B O I
Q P L A S T I C N P Y B S E P C C
B O U N C W A H U J H U G M I Z W

Duration:

6. Forces: *Forces in action*
- Crossword -

Use the clues to fill out the words for the crossword puzzle

Across:
3. Used in car braking system
6. Another name for fulcrum
7. Force applied to a lever
11. Force does work on this in a lever
14. Force reducing length of spring
15. Another name for increased length

Down:
1. Product of mass and 'g' (m x g)
2. A force multiplier (e.g bottle opener)
4. Shape changed due to a force
5. Force-extension graph (Hooke's law)
8. Back to initial shape when force is off
9. New shape even when force is off
10. Another name for force-constant
12. Toothed force multiplier
13. Product of force and ⊥ distance

⊥ used as symbol for perpendicular

Mind-map template: *Forces in action*

Hint: *Write 3 key ideas in the outer boxes*

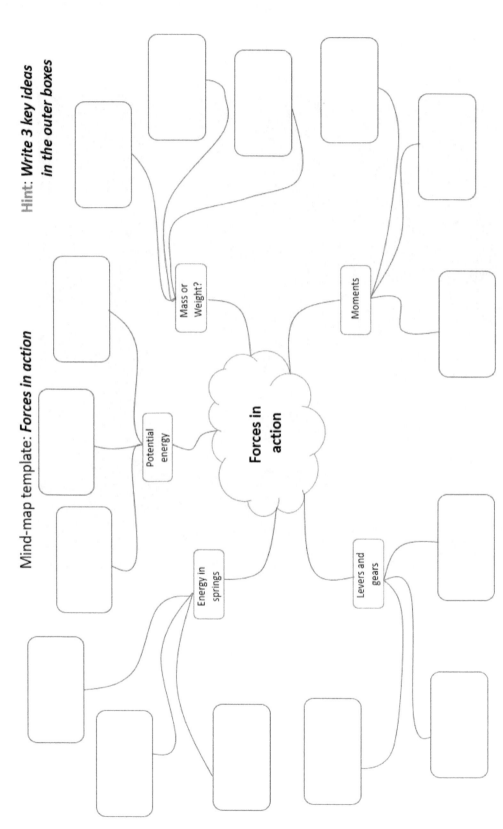

Forces in action

Mass or Weight?

Moments

Potential energy

Energy in springs

Levers and gears

Suggestion: *Use a pencil to complete the mind-map in case you change your mind.*

The Big Picture

The Last three topics form a single unit. Can you see the inter-relationship?

Write (or sketch) below three key ideas that link the three sub-units

We also probe the world around us with big issues/questions

Write down three ideas/questions you would like to follow up on

1.

2.

3.

- Summary -

7. Electricity: *Static and charge*

The effects of static charge are usually fun activities for learners. This unit explores the nature and laws around static and current electricity. It introduces the concept of a field as an area of space where an object experiences a force, electrostatic force in this example. Learners are taken through the journey of science in understanding static charges and discovering that such charges could move through certain materials (as current, the rate of charge flow). It exposes learners to some uses of static electricity, as well as making them aware of the dangers associated with it. Current electricity and associated quantities come next. Learners are reminded of the ability of moving charges to do work; the work done per unit charge as potential difference (loosely as voltage), and the rate of electrical work done as the electrical power. They are then able to use these definitions to calculate the energy consumed around the home, and estimate that used in the schools, and in other places.

7. Electricity: *Static and charge*

Unscramble the key words featured below:

1. tacttar => _ _ _ _ _ _ _

2. docnutroc => _ _ _ _ _ _ _ _ _ _

3. recontel => _ _ _ _ _ _ _ _

4. saturnoil => _ _ _ _ _ _ _ _ _

5. peelr => _ _ _ _ _

6. idlfe => _ _ _ _ _

7. akprs => _ _ _ _ _

8. loomcub => _ _ _ _ _ _ _

9. crtifoni => _ _ _ _ _ _ _ _

10. runcert => _ _ _ _ _ _ _

11. cagehr => _ _ _ _ _

12. actist => _ _ _ _ _ _

7. Electricity: *static and charge*
- Word-search -

Search for the unscrambled key words in the grid of words below. Write the time it took you to find all the words at the bottom of the page.

```
P O T E N T I A L D I F F E R E N C E
I I E I Q J T C E J A Q G R B F U N X
A J A I I U W N V H S R E D K Z B K G
R M O B I N N S O X E P L N I E W R L
S M U D S P S K K R E G O X Q U V O M
B P P L Z U P U V L U K R V T J P T T
A S T E D R Y M L J U E L A Z S S C F
D E F I B R I L L A T O R N H Q H U L
D N M F A K P F V T T H S U J C S D N
E E X P M Y T C I C F O K U Z I F N N
S K I C P N L I Q A B D R V X S D O X
I H N D E M U T H R V X V B X I R C G
L B H X R K G A G T O K P O P T D I H
A P M N E C I T A T S O R T C E L E S
C V H O F I G S V A D T N E R R U C S
O Y T O L J T L L J T Y L S W O A P K
L R M X F U S F Q A I E C Q E U A T U
E V C Y O D O L S B G Q F K Q R S U A
D U U R F R I C T I O N P S K G K N K
```

Duration:

7. Electricity: *static and charge*
- Crossword -

Use the clues to fill out the words for the crossword puzzle

Across:

2. Materials that don't conduct a charge
3. Imparts electrical properties
5. Force between rubbing surfaces
6. Unit of charge
10. Unit of current
13. Applied to restart the heart
14. force between two charged objects
16. force between similar charges
17. Rate of flow of charge

Down:

1. Work done per unit charge
4. Materials that conduct a charge
7. Area where an object feels a force
8. Why sub-atomic particles flow
9. force between dis-similar charges
11. State of charged particles on plastic
12. Sub-atomic particles in current
15. Intense discharge of charges

Mind-map template: *Static and charge*

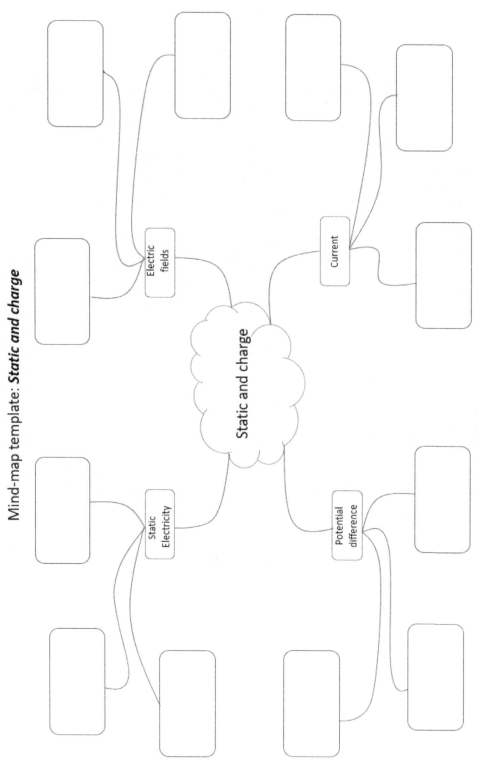

Static and charge

Electric fields

Current

Static Electricity

Potential difference

Suggestion: *Use a pencil to complete the mind-map in case you change your mind.*

8. Electricity: *simple circuits*

From simple circuits learners are exposed to the principles behind the working circuits in the homes, schools, and industries. Through hands-on practical activities, they are exposed to the differences between series and parallel circuits. They revisit the topic of resistance and are reminded that very high resistance could be a desired attribute of electrical heaters. Learners also learn to utilise I-V graphs to determine the nature of electrical components, such as ohmic conductors, filament lamps, and diodes. The last two components are fore-runners of other components that exhibit varying resistance, and to include LDRs and thermistors. The design of control (or smart) circuits is then considered, before delving into quantifying electrical energy consumed by various sectors of the society.

- Anagram -

8. Electricity: *simple circuits*

Unscramble the key words featured below:

1. retnurc => _ _ _ _ _ _ _

2. ancestries => _ _ _ _ _ _ _ _ _ _

3. lapelarl => _ _ _ _ _ _ _ _

4. eiress => _ _ _ _ _ _

5. uriccti => _ _ _ _ _ _ _

6. elmortvet => _ _ _ _ _ _ _ _ _

7. tamerem => _ _ _ _ _ _ _ _

8. smoh => _ _ _ _

9. etnafilm malp => _ _ _ _ _ _ _ _ _ _ _ _ _

10. edodi => _ _ _ _ _

11. ornses => _ _ _ _ _ _

12. shoreimrtt => _ _ _ _ _ _ _ _ _ _

8. Electricity: *simple circuits*
- Word-search -

Search for the unscrambled key words in the grid of words below. Write the time it took you to find all the words at the bottom of the page.

G P F O A B A H K C P S Z M V Y L
F Q V A Q N G E J G A Y L A S Z E
G Q J X V S Z A A M A Y G S Q P L
L H P N U S I T L S R D C J B M L
B X Z T E U U T N E U L E Z O A A
C X T I S R O T S I M R E H T L R
H M R O I Q O I D J R Q G V A T A
G E H X Q U S M N U E G O O X N P
S M A B U T A R V X J S H N X E S
S E M W A O E I B P L B J E E M M
Y S N N F T R T I U C R I C Y A E
N J C S E N J N R U H M G G H L V
C E S M O T V Q R V T B E N A I D
U X M V J R P R Y T K S S P S F I
M A D W W D E V O L T M E T E R O
O V E G H N U H Z N A M E C F Q D
B Y B L T F Z J F P P O S T C J E

Duration:

8. Electricity: *simple circuits*

- *Crossword* -

Use the clues to fill out the words for the crossword puzzle

Across:

3. the unit of resistance
5. used to measure p.d
6. its resistance increases with heating
11. gradient of a V-I graph
12. rate of flow of charges
13. Applied to restart the heart

Down:

1. LDR is used as this to monitor light
2. allows current flow in one direction
4. circuit where supplied p.d is shared
7. used to measure current
8. its resistance decreases with heat
9. circuit where supplied p.d is shared
10. connection of components

Mind-map template: *Simple Circuits*

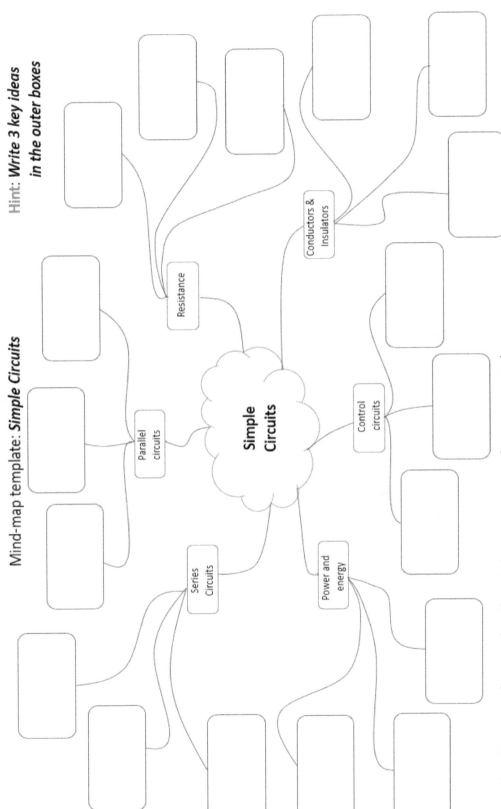

Simple Circuits

Resistance

Conductors & Insulators

Parallel circuits

Control circuits

Series Circuits

Power and energy

The Big Picture

The last two topics form a single unit. Can you see the inter-relationship?

Write (or sketch) below three key ideas that link the two sub-units

We also probe the world around us with big issues/questions

Write down three ideas/questions you would like to follow up on

1.

2.

3.

9. Magnetism: *Magnets and magnetic fields*

Learners revisit the idea of magnetic domains to explain why certain materials are magnetic at room temperature, and others are not. The rules governing the behaviour of magnets and induced magnets are next considered. Next, learners are encouraged to transfer the learning of fields from electrical to the magnetic fields and determine the similarities and differences. Learners learn to map the magnetic field using a compass. They are reminded about the scientifically accepted reasons for the Earth's powerful magnetic field, which serves to shield life on earth from cosmic radiation. Learners are then further challenged to explore magnetic effects of current, in a straight conductor, and later in a solenoid.

- Anagram -

9. Magnetism: *magnets and magnetic fields*

Unscramble the key words featured below:

1. lepo => _ _ _ _

2. acmeting liefd => _ _ _ _ _ _ _ _ _ _ _ _

3. tornh => _ _ _ _ _

4. shout => _ _ _ _ _

5. leste => _ _ _ _ _

6. scamops => _ _ _ _ _ _ _

7. nicedud namget => _ _ _ _ _ _ _ _ _ _ _ _ _

8. mannerpet agmnet => _ _ _ _ _ _ _ _ _ _ _ _ _ _ _

9. noodleis => _ _ _ _ _ _ _ _

10. rruaoa => _ _ _ _ _ _

11. cangemit xflu => _ _ _ _ _ _ _ _ _ _ _ _

12. ismcco ryas => _ _ _ _ _ _ _ _ _ _

9. Magnetism: *magnetic fields and magnets*

- *Word-search* -

Search for the unscrambled key words in the grid of words below. Write the time it took you to find all the words at the bottom of the page.

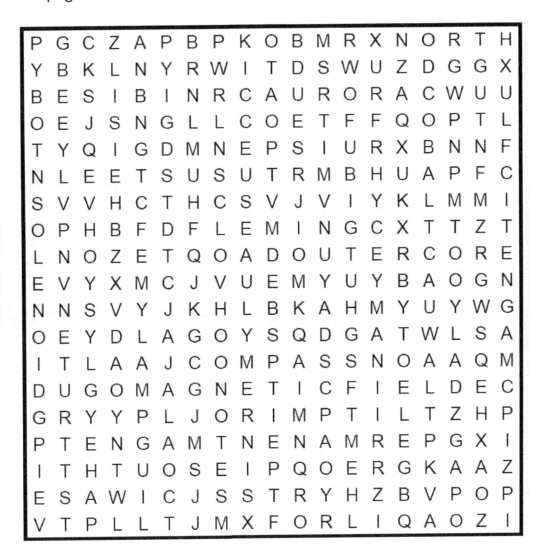

```
P G C Z A P B P K O B M R X N O R T H
Y B K L N Y R W I T D S W U Z D G G X
B E S I B I N R C A U R O R A C W U U
O E J S N G L L C O E T F F Q O P T L
T Y Q I G D M N E P S I U R X B N N F
N L E E T S U S U T R M B H U A P F C
S V V H C T H C S V J V I Y K L M M I
O P H B F D F L E M I N G C X T T Z T
L N O Z E T Q O A D O U T E R C O R E
E V Y X M C J V U E M Y U Y B A O G N
N N S V Y J K H L B K A H M Y U Y W G
O E Y D L A G O Y S Q D G A T W L S A
I T L A A J C O M P A S S N O A A Q M
D U G O M A G N E T I C F I E L D E C
G R Y Y P L J O R I M P T I L T Z H P
P T E N G A M T N E N A M R E P G X I
I T H T U O S E I P Q O E R G K A A Z
E S A W I C J S S T R Y H Z B V P O P
V T P L L T J M X F O R L I Q A O Z I
```

Duration:

9. Magnetism: *magnets and magnetic fields*
- Crossword -

Use the clues to fill out the words for the crossword puzzle

Across:
5. coils of wire
7. magnetic metal at room temperature
8. magnetic, but only attracts
10. unit of magnetic field strength
11. magnetic field lines point into this pole
14. magnetic field lines point away from this pole
15. carbon-containing magnetic alloy
16. molten part of earth, creating current

Down:
1. Left-hand rule named after him
2. Another name for magnetic fields
3. Charged particles from outer space
4. Its magnetic field is always on
6. Earth's deflects charged particles
9. used to map magnetic field
12. made by cosmic rays in atmosphere
13. greatest field strength in a magnet

Mind-map template: *Magnets and Magnetic fields*

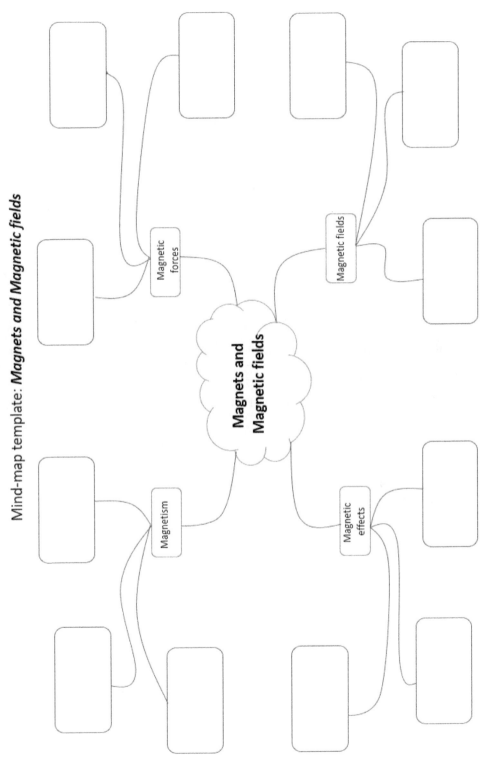

Magnets and Magnetic fields

Magnetic forces

Magnetic fields

Magnetism

Magnetic effects

Suggestion: *Use a pencil to complete the mind-map in case you change your mind.*

- Summary -

10. Magnetism: *Uses of magnetism*

This unit prods learners to deliberate on a few appliances that work on the principle of magnetism. Starting with the force exerted by, and on a current-carrying conductor or loop, learners are exposed to the catapult or motor effect. Subsequently, the learners are encouraged to look around their home to explore and list all the equipment and appliances that depend on motors. In the generator effect we further explore how the interaction between motion and magnetism produces current, which powers our modern lives and industries. Next, we consider the structure of transformers and the principles on which they are based. The learners are then able to appreciate and apply the transformer equation relating the coils ratio to that of the p.d in the primary coil, as well as the secondary coil.

10. Magnetism: *uses of magnetism*

Unscramble the key words featured below:

1. rootm teceff   =>   _ _ _ _ _   _ _ _ _ _ _

2. ulfx dyestin   =>   _ _ _ _   _ _ _ _ _ _ _

3. tunicdoin   =>   _ _ _ _ _ _ _ _ _

4. dicedun cuternr   =>   _ _ _ _ _ _ _   _ _ _ _ _ _ _

5. errantego cefetf   =>   _ _ _ _ _ _ _ _ _   _ _ _ _ _ _

6. tofs niro   =>   _ _ _ _   _ _ _ _

7. macrotumto   =>   _ _ _ _ _ _ _ _ _ _

8. andomy   =>   _ _ _ _ _ _

9. lorentatar   =>   _ _ _ _ _ _ _ _ _ _

10. afroternmrs   =>   _ _ _ _ _ _ _ _ _ _ _

11. deleparksou   =>   _ _ _ _ _ _ _ _ _ _ _

12. ponchomire   =>   _ _ _ _ _ _ _ _ _ _

10. Magnetism: *magnetic fields and magnets*

- Word-search -

Search for the unscrambled key words in the grid of words below. Write the time it took you to find all the words at the bottom of the page.

```
Y B V Z W A B X P W P O F N D L Y H E
U F C L N C B E S V Z Z L O L K C H Q
C P R M Z B L M H C P X H I O O I H B
I F L U X D E N S I T Y K T U Y B J R
L R O T A T U M M O C O E C D A R G E
C P T C E F F E R O T O M U S H A P M
Z K P E Q C R P F J M G J D P O W E R
M S O F T I R O N A J T I N E Q P U O
R I A F G J K C N B W K S I A E L J F
M Z C E T M G Y Q P D T O U K X E E S
D I R R W Q D H I E E I L A E I F J N
T B Y O O B G C K P B L A Q R L G X A
Y U F T O P A O D P H F T P Q E N Z R
Y D Q A O F H O D Q S W I F H K J R T
E T D R W X W O U K H A N B P R U C K
Y U R E H N G Q N J G X G P N T A J D
M J D N K Z A L T E R N A T O R R G Z
D D Y E I N D U C E D C U R R E N T M
K V C G A N S P G S I W P H C D W Q T
```

Duration:

10. Magnetism: *magnets and magnetic*
- Crossword -

Use the clues to fill out the words for the crossword puzzle

Across:
1. another name for the field strength
5. in it moving coil generates current
7. material used as core of **2 Down**
10. flowing charges in secondary coils
12. **13 Across** produces d.c. current
13. motion and magnet create current
15. vibrates to move coil in **5 Across**
17. device where current moves **15 Across**

Down:
2. Device that changes output p.d
3. Output p.d is same as input in **2 Down**
4. Also called catapult effect
6. This split ring ensures d.c. current
8. Process of output current in **2 Down**
9. **13 Across** produces a.c current
11. A type of **2 Down,** output p.d is less
14. current produced in **7 Across**
16. ≈ equal for both coils in **2 Down**

Mind-map template: *Uses of magnetism*

Hint: *Write 3 key ideas in the outer boxes*

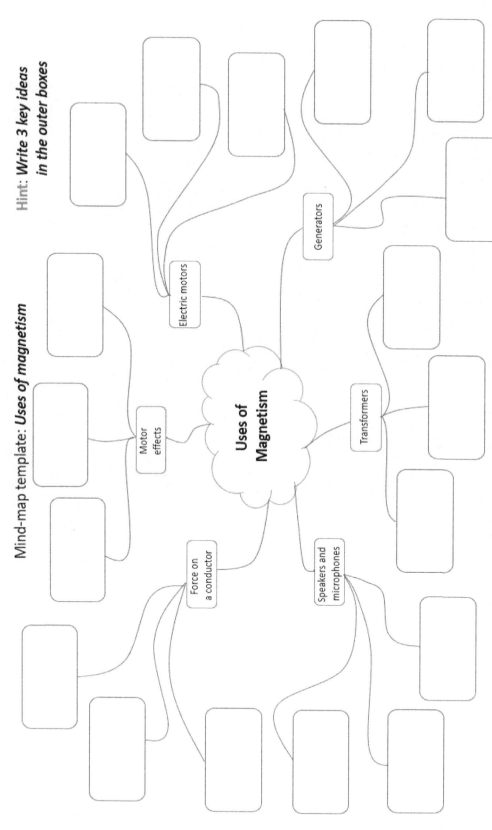

Suggestion: *Use a pencil to complete the mind-map in case you change your mind.*

The Big Picture

The last two topics form a single unit. Can you see the inter-relationship?

Write (or sketch) below three key ideas that link the two sub-units

We also probe the world around us with big issues/questions

Write down three ideas/questions you would like to follow up on

1.

2.

3.

Answers

Answers

1. Matter: *The particle model*

Anagrams

1. atom

2. atomic model

3. anomalous

4. density

5. energy level

6. neutral

7. nucleus

8. orbit

9. shell

10. particle model

11. ratio

12. relative

13. mass

14. charge

15. volume

Answers

1. Matter: *The particle model (Word-search & Crossword)*

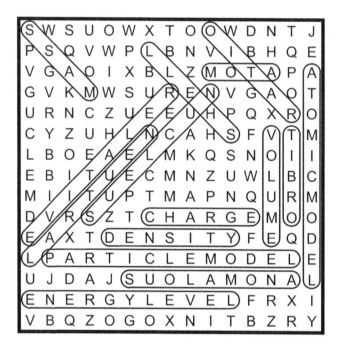

Word-search
1. atom
2. atomic model
3. anomalous
4. density
5. energy level
6. neutral
7. nucleus
8. orbit
9. shell
10. particle model
11. ratio
12. relative
13. mass
14. charge
15. volume

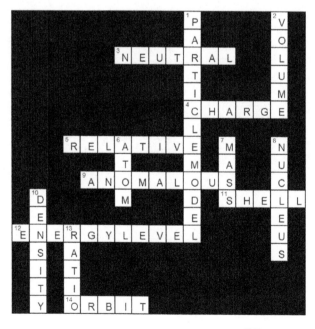

Across: 3 Neutral,
4 Charge, **5** Relative,
9 Anomalous, **11** Shell,
12 Energy level, **14** Orbit.

Down: 1 Particle model,
2 Volume, **6** Atom,
7 Mass, **8** Nucleus,
10 Density, **13** Ratio.

Answers

2. Matter: *Changes of state*

Anagrams

1. absolute scale
2. boiling
3. bonds (Intermolecular)
4. change of state
5. condensation
6. conservation (of mass)
7. energy transfer
8. evaporation
9. freezing
10. Intermolecular
11. internal energy
12. kinetic energy
13. melting
14. potential energy
15. (specific) heat (capacity)
16. (specific) latent (heat)
17. sublimation
18. temperature

Answers

2. Matter: *Changes of state (Word-search & Crossword)*

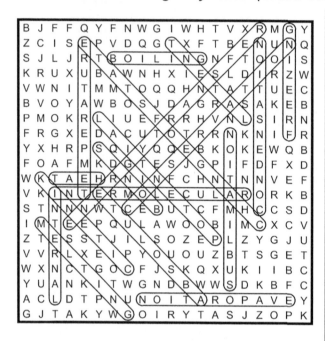

Word-search

1. Absolute (scale)
2. Boiling
3. Bonds (intermolecular)
4. Change of state
5. Condensation
6. Conservation (of mass)
7. Energy transfer
8. Evaporation
9. Freezing
10. Heat (specific heat capacity)
11. Intermolecular
12. Internal (energy)
13. Kinetic (energy)
14. Latent (specific latent heat)
15. Melting
16. Potential (energy)
17. Sublimation
18. Temperature

Across: 6 Potential, **8** Latent, **9** Melting, **13** Change of state, **15** Sublimation, **16** Condensation, **17** Kinetic.

Down: 1 Freezing, **2** Evaporation, **3** Boiling, **4** Internal, **5** Heat, **7** Intermolecular, **10** Energy transfer, **11** Bonds, **12** Conservation, **14** Absolute.

Answers

3. **Matter:** *Pressure*

Anagrams

1. atmospheric pressure
2. gas pressure
3. compressed
4. incompressibility
5. inversely proportional
6. normal (pressure)
7. pascal (Pa)
8. randomly
9. temperature
10. upthrust
11. volume

Answers

3. Matter: *Changes of state (Word-search & Crossword)*

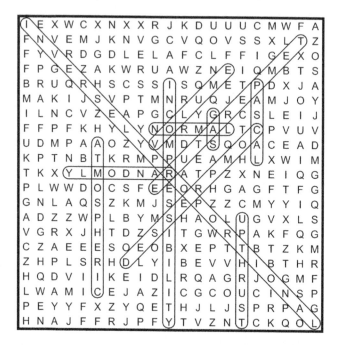

Word-search

1. Atmospheric
2. Compressed
3. Gas
4. Incompressibility
5. Inversely proportional
6. Normal
7. Pascal
8. Randomly
9. Temperature
10. Upthrust
11. Volume

Across: 2 Upthrust, **6** Pascal, **7** Temperature, **10** Normal, **11** Inversely proportional.

Down: 1 Atmospheric, **3** Incompressibility, **4** Gas, **5** Volume, **8** Randomly, **9** Compressed.

Answers

4. Forces: *Motion*

Anagram

1. Acceleration
2. Air-resistance
3. Average-speed
4. Deceleration
5. Displacement
6. Distance
7. Gradient
8. Rate of change
9. Scalar
10. Speed
11. Tangent
12. Uniform motion
13. Velocity

Answers

4. Matter: *Motion (Word-search and Crossword)*

```
W V L A U Y J P M S L N H T H J P Y
D D Y C K K F V G C I P P K H O W A
Y R W C R R C Y J A R G C E E F L T
S A P E K S Q W X L A J N N K K T G
F D J L T N E G N A T D A N C S V Z
L D E E P S E G A R E V A J U A T N
A Q M R F F Z V U C O L I Z T U Z K
G C R A G S Y D E B F E R X M A T I
C R I T T X D L F Y C O R H P H N L
L M M I S Q E N B N H W E K F L B H
U Z B O Q R V E A L A Y S P E E D T
S I O N A F A T J O N Q I L K W U N
L U P T G M S D W P G I S M M S F E
E H I D D I X E U O E Z T P W P R I
T O Y E D T T N E M E C A L P S I D
N N O I T O M M R O F I N U D K J A
F W Q F U U Q M S Y W C C I T F R R
Y D P Z K W Y T I C O L E V U C Q G
```

Word-search

1. Acceleration
2. Air-resistance
3. Average-speed
4. Deceleration
5. Displacement
6. Distance
7. Gradient
8. Rate of change
9. Scalar
10. Speed
11. Tangent
12. Uniform motion
13. Vector
14. Velocity

Across: 4 Speed, **6** Acceleration, **7** Deceleration, **8** Distance, **9** Tangent, **12** Air-resistance, **13** Uniform motion.

Down: 1 Rate of change, **2** Displacement, **3** Scalar, **5** Velocity, **10** Gradient, **11** Vector.

Answers

5. Motion : Newton's Laws of motion

Anagrams

1. crumple zone
2. equilibrium
3. force
4. balanced
5. component
6. contact
7. resultant
8. free-body
9. inertia
10. momentum
11. power
12. work

Answers

5. Motion: *Newton's Laws of motion (Word-search and Crossword)*

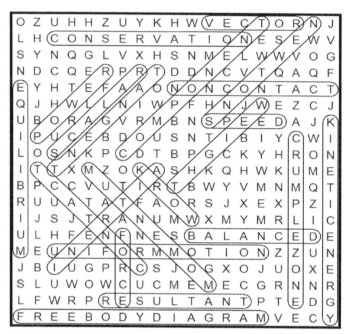

Word-search

1. balanced
2. component
3. conservation
4. contact
5. crumple zone
6. equilibrium
7. force
8. free-body diagram
9. inertia
10. kinetic energy
11. momentum
12. newtons
13. non-contact
14. power
15. resultant
16. scalar
17. speed
18. tangent
19. uniform motion
20. vector
21. weight
22. work

Across: 1 Conservation, 3 Equilibrium, 5 Scalar, 6 Tangent, 7 Power, 8 Non-contact, 11 Momentum, 15 Newtons, 17 Inertia, 18 Vector, 19 Kinetic energy, 20 Uniform motion.

Down: 1 Crumple zone, 2 Weight, 4 Free-body diagram, 9 Component, 10 Balanced, 12 Resultant, 13 Force, 14 Speed, 16 Work.

Answers

6. Motion : Forces in action

Anagrams

1. compression
2. deformation
3. elastic
4. plastic
5. extension
6. gear
7. lever
8. hydraulic
9. linear
10. moment
11. pivot
12. stiffness

Answers

6. Motion: *Forces in action (Word-search & Crossword)*

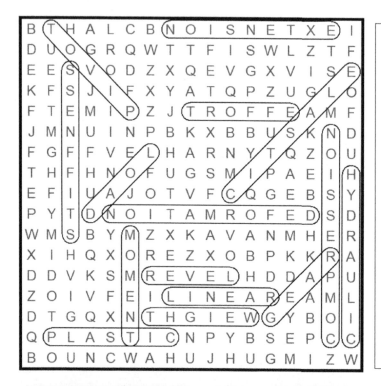

Word-search

1. compression
2. deformation
3. effort
4. elastic
5. plastic
6. extension
7. gear
8. lever
9. load
10. hydraulic
11. linear
12. moment
13. pivot
14. stiffness
15. weight

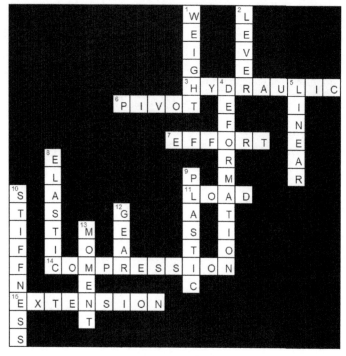

Across: 3 Hydraulic,
6 Pivot, **7** Effort,
11 Load,
14 Compression,
15 Extension.

Down: 1 Weight,
2 Lever, **4** Deformation,
5 Linear, **8** Elastic,
9 Plastic, **10** Stiffness,
12 Gear, **13** Moment.

Answers

7. Electricity : Static and charge

Anagrams

1. attract
2. conductor
3. electron
4. insulator
5. repel
6. field
7. spark
8. coulomb
9. friction
10. current
11. charge
12. static

Answers

7. Electricity: *Static and charge (Word-search & Crossword)*

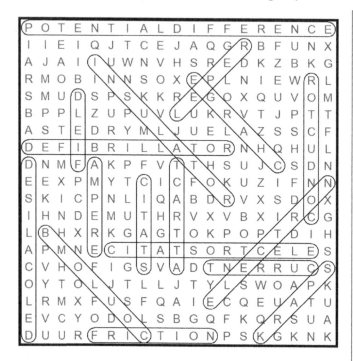

Word-search

1. ampere
2. attract
3. conductor
4. electron
5. insulator
6. repel
7. field
8. spark
9. coulomb
10. electrostatic
11. defibrillator
12. friction
13. delocalised
14. current
15. potential difference
16. charge
17. static

Across: 2 Insulator,
3 Charge, **5** Friction,
6 Coulomb, **10** Ampere,
13 Defibrillator,
14 Electrostatic,
16 Repel, **17** Current.

Down: 1 Potential
difference, **4** Conductor,
7 Field, **8** Delocalised,
9 Attract, **11** Static,
12 Electron, **15** Spark.

Answers

8. Electricity: *Simple circuits*

Anagrams

1. current
2. resistance
3. parallel
4. series
5. circuit
6. voltmeter
7. ammeter
8. ohms
9. filament lamp
10. Diode
11. sensor
12. thermistor

Answers

8. Electricity: *Simple circuits (Word-search & Crossword)*

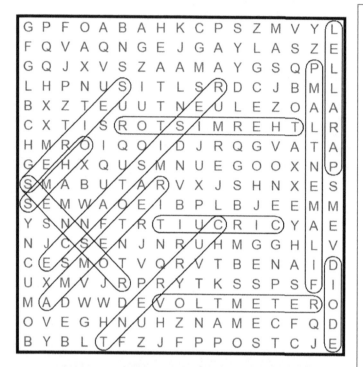

Word-search

1. conductor
2. insulator
3. current
4. potential difference
5. resistance
6. parallel circuit
7. series circuit
8. voltmeter
9. ammeter
10. ohm-meter
11. filament lamp
12. Diode
13. LDR
14. sensors
15. thermistor
16. power
17. energy

Across: 3 Ohms,
5 Voltmeter, **6** Filament lamp, **11** Resistance,
12 Current.

Down: 1 Sensor, **2** Diode,
4 Series, **7** Ammeter,
8 Thermistor, **9** Parallel,
10 Circuit.

Answers

9. Magnetism: *Magnets and magnetic fields*

Anagrams

1. pole
2. magnetic field
3. north
4. south
5. steel
6. compass
7. induced magnet
8. permanent magnet
9. solenoid
10. aurora
11. magnetic flux
12. cosmic rays

Answers

9. Magnetism: *Magnets and magnetic fields* *(Word-search & Crossword)*

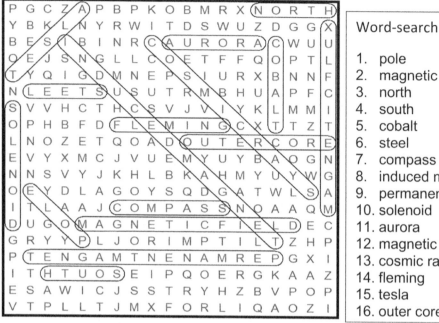

Word-search

1. pole
2. magnetic field
3. north
4. south
5. cobalt
6. steel
7. compass
8. induced magnet
9. permanent magnet
10. solenoid
11. aurora
12. magnetic flux
13. cosmic rays
14. fleming
15. tesla
16. outer core

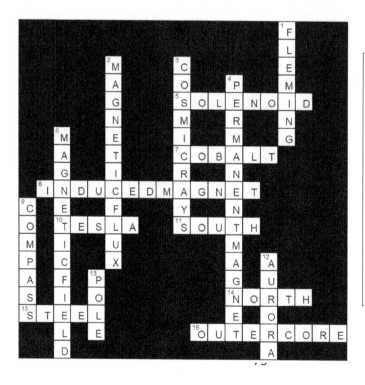

Across: 5 Solenoid,
7 Cobalt, **8** Induced
magnet, **10** Tesla,
11 South, **14** North,
15 Steel, **16** Outer core.

Down: 1 Fleming,
2 Magnetic flux,
3 Cosmic rays,
4 Permanent magnet,
6 Magnetic field,
9 Compass, **12** Aurora,
13 Pole.

Answers

10. Magnetism: *Uses of magnetism*

Anagrams

1. Motor effect
2. Flux density
3. induction
4. Induced current
5. Generator effect
6. soft iron
7. Commutator
8. Dynamo
9. Alternator
10. Transformer
11. Loudspeaker
12. microphone

10. Magnetism: *Magnets and magnetic fields (Word-search & Crossword)*

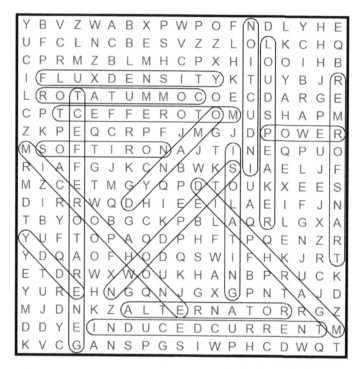

Word-search

1. Motor effect
2. Flux density
3. induction
4. Induced current
5. Generator effect
6. soft iron
7. Commutator
8. Dynamo
9. Alternator
10. Transformer
11. Loudspeaker
12. microphone
13. diaphragm
14. isolating
15. step-down
16. eddy
17. power

Across: 1 Flux density,
5 Microphone, **7** Soft iron, **10** Induced current, **12** Dynamo, **13** Generator effect, **15** Diaphragm, **17** Loudspeaker.

Down: 2 Transformer, **3** Isolating, **4** Motor effect, **6** Commutator, **8** Induction, **9** Alternator, **11** Step-down, **14** Eddy, **16** Power.